Ángeles García Aliaga

# PET-CT scan protocol and dosimetry evaluation

Ángeles García Aliaga

# PET-CT scan protocol and dosimetry evaluation

## Evaluation of the protocol and dosimetry of PET-CT scanning with [18F]F-FDG performed at HGUSL.

ScienciaScripts

**Imprint**
Any brand names and product names mentioned in this book are subject to trademark, brand or patent protection and are trademarks or registered trademarks of their respective holders. The use of brand names, product names, common names, trade names, product descriptions etc. even without a particular marking in this work is in no way to be construed to mean that such names may be regarded as unrestricted in respect of trademark and brand protection legislation and could thus be used by anyone.

Cover image: www.ingimage.com

This book is a translation from the original published under ISBN 978-3-639-55782-4.

Publisher:
Sciencia Scripts
is a trademark of
Dodo Books Indian Ocean Ltd. and OmniScriptum S.R.L publishing group

120 High Road, East Finchley, London, N2 9ED, United Kingdom
Str. Armeneasca 28/1, office 1, Chisinau MD-2012, Republic of Moldova, Europe
Printed at: see last page
**ISBN: 978-620-7-65682-0**

SUMMARY

Patient safety is a fundamental pillar in the field of medicine and pharmacy. As diagnostic techniques and treatments advance, it is essential to ensure that the clinical benefits far outweigh the associated risks in any field, but even more so in medical specialties linked to ionizing radiation exposure. In this sense, the reduction of dosimetry in medical tests has become an area of great interest, since the aim is to minimize patient exposure to ionizing radiation without compromising the quality of the clinical results obtained. Nuclear medicine together with radiopharmacy are health specialties that in their diagnostic branch use radiopharmaceuticals for metabolic and/or molecular imaging. The present work focuses on the impact from a patient safety point of view of modifying the dosage of the radiopharmaceutical [18F]F-fludesoxyglucose commonly used in PET-CT scans. The main objective of this study is to evaluate the effectiveness of changing the dosage of this radiopharmaceutical from a multidisciplinary approach to radioprotection and patient safety without compromising the quality of the images and the clinical results obtained.

**INDEX**

# 1. INTRODUCTION

## 1.1 PET-CT scan

Nuclear medicine is a medical specialty that uses diagnostic imaging techniques to obtain images through the administration of radioactive substances called radiopharmaceuticals. The techniques used in nuclear medicine are based on the principles of nuclear physics, radiopharmacy and clinical medicine to obtain images of metabolic processes and/or molecular structures of the organism under study. The fundamentals of nuclear medicine could be summarized as follows:

- Radiopharmaceuticals: Radiopharmaceuticals are drugs used in nuclear medicine, composed of a radioactive isotope and a biologically active molecule. These radiopharmaceuticals emit radiation, which can be detected and used to obtain information about the structure and function of the target organ or tissue.
- Diagnosis: Nuclear medicine provides functional and molecular imaging for diagnosis, allowing detailed information to be obtained on metabolism, perfusion and cellular activity in different organs and tissues. The most common scans in nuclear medicine are scintigraphy, *Single-Photon Emission Computed Tomography* (SPECT) and *Positron Emission Tomography* (PET).
- Therapy: Nuclear medicine also makes it possible to treat certain pathologies. In this case, therapeutic radiopharmaceuticals are administered that emit radiation directly to the target cells, with the aim of destroying them or reducing their activity.
- Monitoring and response to treatment: the ability of radiopharmaceuticals to track biological and metabolic processes allows assessment of the effectiveness of therapies and close monitoring to detect any changes or progression of the disease.
- Safety: to ensure the safety of this discipline it is necessary to apply radiation safety protocols to protect the patient, healthcare personnel and the environment. It is the responsibility of specialists in nuclear medicine, radiopharmacy and hospital

radiophysics to establish protocols for safe handling of radiopharmaceuticals and proper management of ionizing radiation.

Within this field of medicine, it is worth highlighting PET-CT scanning, a non-invasive diagnostic imaging scan that combines two different techniques to obtain detailed images of the pathology to be studied. On the one hand, a computed tomography (CT) scan is performed in which multiple X-ray images of the body are taken from different angles to create a detailed image of the internal structures. Subsequently, PET is performed which detects the gamma rays emitted by the radiopharmaceutical administered according to the indication of the test and creates a three-dimensional image of the metabolic and/or molecular activity of the organism. The fusion of these images allows nuclear medicine

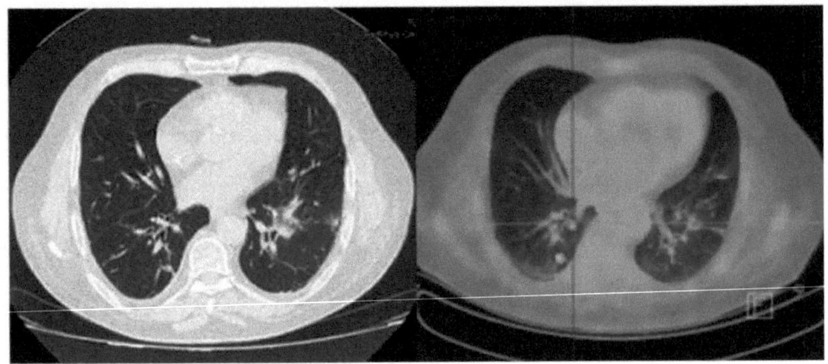

specialists to see both the metabolic activity and the anatomical structure of organs and tissues (**Figure 1**).

Malignant pulmonary nodule with [18F]F-fludesoxyglucose uptake in PET-CT image (right) but undetectable in CT (left). Taken from Lopez Lopez et al. (1)

The role of radiopharmacy, the multidisciplinary and hospital-based health specialty that studies the pharmaceutical, chemical, biochemical, biochemical, biological and physical aspects of radiopharmaceuticals, is essential for carrying out these explorations. This specialty also applies this knowledge to the processes of design, production, preparation, quality control and dispensing of radiopharmaceuticals, both in healthcare (diagnostic and therapeutic) and in research. Radiopharmacists, healthcare professionals specializing in radiopharmacy with specialized training via QIR or FIR in specific knowledge of radiopharmaceuticals, are responsible for the proper use of radiopharmaceuticals through the appropriate selection, custody and management of these, with the aim of achieving the optimum ratio of quality, safety and cost-effectiveness, in accordance with current legislation. The responsibility, supervision and control of the proper use of radiopharmaceuticals corresponds legally to the radiopharmacy specialist, as well as the preparation of PET radiopharmaceuticals.

### 1.2 PET radiopharmaceuticals

Radionuclides of artificial origin emitting ionizing radiation are used in nuclear medicine. They are obtained by bombarding nuclei of stable atoms with subatomic particles (neutrons, protons, etc.) causing nuclear reactions to convert stable nuclei into unstable (radioactive) ones. Devices and methods used to produce radionuclides include: nuclear reactors, particle accelerators such as cyclotrons and generators. In this case, obtaining the radionuclides used in PET is mainly carried out by means of cyclotrons or the use of generators.

The decay of a positron-emitting radionuclide occurs according to the following general scheme: ${}^{A}_{Z}X_{N} \rightarrow {}^{A}_{Z-1}Y_{N+1} + \beta^{+} + \nu$

As the β+ radiation emitting radioisotope injected into the patient decays, it emits positrons. Each of these may collide with a cortical electron from the tissue in which the radiopharmaceutical is present and annihilation occurs. This phenomenon results in the emission of two 511keV gamma rays in the same direction and diametrically opposite.

The rationale behind the use of PET radiopharmaceuticals labeled with positron emitting isotopes is that they allow in vivo visualization of various physiological or pathophysiological processes. In this way, it is possible to monitor the temporal evolution of the regional distribution of the concentration of a radiopharmaceutical after administration of the labeled compound. The number of PET radiopharmaceuticals available to date is very large.

However, most of them have been used in research studies and only a few have been introduced into routine clinical practice. Such is the case of [18F]F-fludesoxyglucose, hereafter [18F]F-FDG, the most widely used PET radiopharmaceutical that allows the study of cellular glucose metabolism. The success of this radiopharmaceutical is due in part to the fact that it meets the ideal characteristics that a PET radiopharmaceutical should have: easy penetration into the target tissue, low nonspecific uptake, high affinity for its binding site, sufficiently slow dissociation from the binding site to detect such binding after elimination of the nonspecifically bound compound (metabolic trapping) and low metabolization to facilitate image acquisition (2).

### 1.3 [18F]F-FDG

18F]F-FDG is undoubtedly the most important PET radiopharmaceutical. This is due not only to its application in the study of a wide range of pathologies, but also to its metabolic characteristics and the speed of its synthesis. Both glucose and [18F]F-FDG cross the blood-brain barrier and easily enter cells, although this incorporation step is slightly faster in the case of the fluorinated analog. After entry into the cell, both compounds initiate the glycolytic pathway. However, [18F]F-FDG undergoes only the first step of the glycolytic pathway because the resulting compound ([18F]FDG-6-P) is a substrate incompatible with phosphoglucose isomerase and cannot be metabolized (metabolic trapping).

The most widespread use of [18F]F-FDG is the study of tumor pathology thanks to its role as a marker of cellular glycogen metabolism. The high concentration of [18F]F-FDG inside tumor cells is a reflection of their increased glycogen metabolism in order to maintain a high rate of growth and/or proliferation. The need for energy in the form of ATP for these anabolic processes translates into increased glucose uptake. Therefore, the use of [18F]F-FDG in oncology is based on the observation that tumor cells show increased glycolysis.

According to data sheet of the radiopharmaceutical GLUSCAN 600 MBq/mL solution for injection (one of the commercial presentations of the [18F]F-FDG), this drug is indicated for (3):

- Oncology: diagnosis, staging, monitoring of response to treatment and detection of recurrence in various tumors.
- Cardiology: myocardial viability assessment.
- Neurology: diagnosis of interictal hypometabolism.
- Infectious or inflammatory diseases: diagnosis of structures with activated leukocytes, fever of unknown origin, among others.

As for the posology for adults and elderly patients, an activity of 100 to 400 MBq is recommended for an adult weighing 70 kg. This activity should be adjusted according to the patient's body weight, the type of camera used and the mode of image acquisition, administered by direct intravenous injection. (3). However, the *Society of Nuclear Medicine and Molecular Imaging* (SNMMI) and the *European Association of Nuclear Medicine* (EANM) recommend a dosage of 3.7-5.2 MBq/kg of [18F]F-FDG for body PET-CT scans depending on the type of camera used and imaging mode of acquisition (4).

### 1.4 Dosimetry and radiation protection

Dosimetry is a scientific subspecialty, in the field of health physics and medical physics, focused on the calculation of internal and external doses of radiation. This dosimetry is measurable in Gray (Gy) or Sievert (Sv) according to the calculation used. The dose absorbed in tissues and matter as a result of exposure to ionizing radiation, both directly and indirectly, is measured in Sievert (Sv), the ionizing radiation dose equivalence unit of the International System of Units (SI). The internal dosimetry of absorbed dose per unit of activity administered can be calculated according to ICRP (International Commission on Radiological Protection) publication No. 106 (**Table 1**). (5).

| ORGAN | DOSE ABSORBED PER UNIT OF ACTIVITY ADMINISTERED (mGy/MBq) | | | | |
|---|---|---|---|---|---|
| | Adult | 15 years | 10 years | 5 years | 1 year |
| Adrenal glands | 0,012 | 0,016 | 0,024 | 0,039 | 0,071 |
| Bladder | 0,13 | 0,16 | 0,25 | 0,34 | 0,47 |
| Bone surfaces | 0,011 | 0,016 | 0,022 | 0,034 | 0,064 |
| Brain | 0,038 | 0,039 | 0,041 | 0,046 | 0,063 |
| Mamas | 0,0088 | 0,011 | 0,018 | 0,029 | 0,056 |
| Gallbladder | 0,013 | 0,016 | 0,024 | 0,037 | 0,070 |
| Gastrointestinal tract | | | | | |
| Stomach | 0,011 | 0,014 | 0,022 | 0,035 | 0,067 |
| Small intestine | 0,012 | 0,016 | 0,025 | 0,040 | 0,073 |
| Colon | 0,013 | 0,016 | 0,025 | 0,039 | 0,070 |
| Ascending large intestine | 0,012 | 0,015 | 0,024 | 0,038 | 0,070 |
| Descending large intestine | 0,014 | 0,017 | 0,027 | 0,041 | 0,070 |
| Heart | 0,067 | 0,087 | 0,13 | 0,21 | 0,38 |
| Kidneys | 0,017 | 0,021 | 0,029 | 0,045 | 0,078 |
| Liver | 0,021 | 0,028 | 0,042 | 0,063 | 0,12 |
| Lungs | 0,020 | 0,029 | 0,041 | 0,062 | 0,12 |
| Muscles | 0,010 | 0,013 | 0,020 | 0,033 | 0,062 |
| Esophagus | 0,012 | 0,015 | 0,022 | 0,035 | 0,066 |
| Ovaries | 0,014 | 0,018 | 0,027 | 0,043 | 0,076 |
| Pancreas | 0,013 | 0,016 | 0,026 | 0,040 | 0,076 |
| Red bone marrow | 0,011 | 0,014 | 0,021 | 0,032 | 0,059 |
| Skin | 0,0078 | 0,0096 | 0,015 | 0,026 | 0,050 |
| Spleen | 0,011 | 0,014 | 0,021 | 0,035 | 0,066 |
| Testicles | 0,011 | 0,014 | 0,024 | 0,037 | 0,066 |
| Thyroid | 0,010 | 0,013 | 0,021 | 0,034 | 0,065 |
| Uterus | 0,018 | 0,022 | 0,036 | 0,054 | 0,090 |
| Rest of the organism | 0,012 | 0,015 | 0,024 | 0,038 | 0,064 |
| Effective Dose (mSv/MBq) | 0,019 | 0,024 | 0,037 | 0,056 | 0,095 |

**Table 1**. Internal dosimetry of [18F]F-FDG, dosimetry calculated as absorbed dose per unit of activity administered (mGy/MBq). Taken from the Technical Data Sheet of GLUSCAN 600 MBq/mL solution for injection. (3).

On the other hand, the basis of radiation protection is the protection of individuals, their descendants and mankind as a whole from the risks arising from those activities which, due to the equipment or materials they use, involve exposure to ionizing radiation. There are two types of biological effects from exposure to ionizing radiation. The effects that occur with certainty when a given value of received dose is exceeded (deterministic) and those that have an increasing probability of occurrence as the dose increases (stochastic). In this sense, the ICRP considers that the main objective of radiation protection is to avoid the occurrence of deterministic biological effects and to limit as much as possible the probability of occurrence of stochastic ones.

The three basic principles of the current ICRP recommendations are:

- Justification: No practice involving exposure to ionizing radiation should be adopted if its introduction does not produce a positive net benefit.
- Optimization or ALARA principle: ALARA stands for "*As Low As Reasonably Achievable*". All radiation exposures should be kept as low as reasonably achievable. All radiation doses involve some risk, so it is not sufficient to comply with the dose limits that are set in national regulations. Doses should be further reduced, whenever reasonably possible.
- Dose limit: radiation doses received by people must not exceed the limits established in national regulations. The dose limits established in Spanish legislation guarantee that people are not exposed to an unacceptable level of risk.

The Nuclear Safety Council (CSN) is the Spanish institution whose purpose is to ensure nuclear safety and radiological and environmental protection. It considers that patients undergoing medical examinations with ionizing radiation should be given special consideration from the point of view of radiological protection, since this exposure should be of great diagnostic or therapeutic benefit compared to the possible harm they may cause. In addition, diagnostic procedures should be optimized in order to obtain an adequate diagnostic image with the lowest possible dose. (6).

## 2. OBJECTIVES

The Law 29/2006, of July 26, 2006, on guarantees and rational use of drugs and medical devices , considers radiopharmaceuticals in its chapter V as special drugs with their own specific regime, in which the specialist in radiopharmacy is the competent professional for their preparation in duly authorized radiopharmacy units. (7). Furthermore, in the Order SCO/2733/2007, of September 4, which approves and publishes the training program for the specialty of Radiopharmacy , it is stated that the radiopharmacy specialist is responsible for the proper use of radiopharmaceuticals through their adequate selection, custody and management, in order to achieve optimal use with quality, safety and cost-effectiveness, in accordance with the principles of correct radiopharmaceutical preparation and the legislation in force. (8).

On the other hand, the Royal Decree 673/2023, of July 18, establishing the quality and safety criteria for nuclear medicine care units determines that the nuclear medicine specialist is responsible for prescribing the radiopharmaceutical to be used and the activity to be administered (applicable to diagnostic and therapeutic radiopharmaceuticals). (9).

In view of the above, it should be noted that the nuclear medicine department of the Hospital General Universitario Santa Lucía does not prescribe PET radiopharmaceuticals. PET-CT scans with [18F]F-FDG are performed using a generic dosage of 360-400 MBq per patient, which does not take into account the technical specifications of the camera used, nor does it adjust to the patient's weight or body surface area. This protocol implies an increase in the dosimetry of PET-CT scanning with [18F]F-FDG in most patients; for this reason, the main objective of this work is to modify this protocol through multidisciplinary advice and to check the quality of the scans. As a secondary objective, it is proposed to make a registry of all the medical prescriptions of PET radiopharmaceuticals according to the indicated PET-CT scan.

## MATERIAL AND METHODS

### 3.1 Scope and function

The Hospital General Universitario Santa Lucía (HGUSL), together with the Hospital General Universitario Santa María del Rosell, forms the Hospital Complex of Cartagena. This hospital is part of the health area II (Cartagena), which includes the municipalities of Cartagena, La Unión, Fuente Álamo and Mazarrón. It belongs to the Servicio Murciano de Salud (SMS), the body in charge of the public health care system in the Spanish autonomous community of the Region of Murcia. The SMS currently has two Nuclear Medicine Services: one at the Hospital Clínico Universitario Virgen de la Arrixaca (HCUVA), inaugurated in 1976, and the other at the HGUSL, inaugurated in 2011. The nuclear medicine service of the HCUVA, a reference hospital in the region, has 2 SPETC-CT gamma cameras, 1 SPETC gamma camera and 1 PET-CT. However, the nuclear medicine department of the HGUSL only has one SPETC-CT gamma camera and one PET-CT. The nuclear medicine service of HGUSL is located on the first floor (P0) of block 5 of this hospital and the PET-CT facilities are located in the basement (P-1) of the same block.

At the time of writing, the PET-CT equipment available in the Region of Murcia is 2 PET-CT chambers for almost 1.5 million inhabitants (this translates into 750,000 inhabitants per chamber). This situation compared with the province of Alicante (1.86 million inhabitants for 5 PET-CT chambers, which means 372,000 inhabitants per chamber) leads to an increase in waiting lists for these tests, to an overload of care and in some cases to work *burnout.* Due to this fatigue, colleagues in the nuclear medicine department had never considered changing the protocol for PET-CT scanning with [18F]F-FDG or the use of medical prescriptions for diagnostic tests.

### 3.2 Intervention

#### 3.2.1 Identification of opportunities for improvement

The main function of the radiopharmacy unit, located within the nuclear medicine service of the HGUSL, is the proper use of radiopharmaceuticals through their selection, custody and management. The final responsibility for ensuring that this unit complies with all current legislation lies directly with the radiopharmacy specialist, whose mission in this work has been to investigate, analyze protocols and propose improvements, if possible, that will benefit the patients of the nuclear medicine service of the HGUSL.

### 3.2.1 Design of the quality assessment study

Experimental research with a before-after design to evaluate by means of a qualitative and quantitative approach the improvement cycle carried out in the evaluation of the protocol and dosimetry of PET-CT scanning with [18F]F-FDG.

- Title of the study
  - " Evaluation of the protocol and dosimetry of PET-CT scanning with [18F]F-FDG performed at the nuclear medicine department of HGSL "
- Definition of the problem or opportunity for improvement
  - Improvement and adaptation of PET-CT scanning protocols with [18F]F-FDG to current legislation.
- Analysis performed
  - The causes that can influence the identified quality problem were analyzed using a cause-effect diagram.

### 3.2.2 Cause-effect diagram

Root cause analysis (RCA), also known as fishbone because of its graphical appearance or Ishikawa diagram, is a tool that allows to break down a quality problem into identifiable and manageable elements for intervention considered as potential causes of the problem studied. This makes it possible to identify all the different causes that can influence the quality problem under study or to detect problem areas globally. Once the causes are known, it is useful to reflect on the level of knowledge of the problem as a whole and whether it is necessary to extend the degree of knowledge of these potential causes. Finally, the quantification of these causes can be performed to start the experimental part of the cycle.

The diagram represents the relationship between a quality problem, in this case the quality problems associated with the "[18F]F-FDG PET-CT scan protocol" (effect), and its potential causes, as shown in the following figure (**Figure 2**).

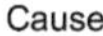

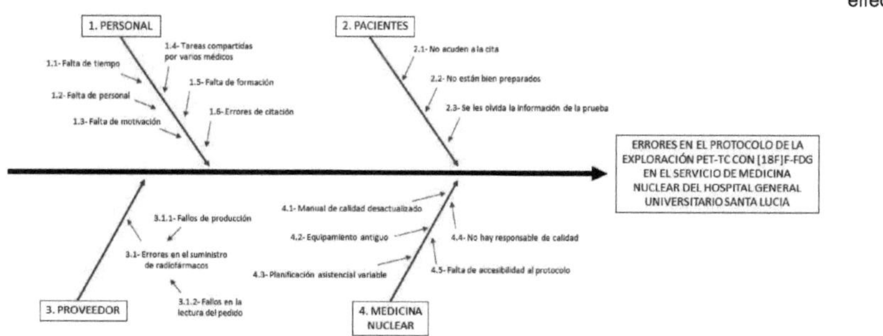


Cause and effect diagram to analyze the opportunity for improvement: improvement and adaptation of PET-CT scanning protocols with [18F]F-FDG to current legislation.

### 3.2.3 Definition of causes presented in terms of their operational translation for the continuity of the improvement cycle

Depending on the causes that influence an identified quality problem, these may be:

- Non-modifiable causes: those that cannot be intervened, but it is important to take them into account since they can lead to situations to be taken into consideration.
- Hypothetical modifiable causes: it is perceived that there may be a relationship with the quality problem, they can be modified with available resources and knowledge.
- Scientifically evidenced modifiable causes not quantified: the frequency with which they occur is not known, determining the existing level of quality.
- Scientifically evidenced and quantified modifiable causes: those in which the frequency with which they occur is known, and which should be addressed, but are not being addressed.

According to these definitions, the causes influencing the quality problem identified in this study can be categorized as shown in the following table (**Table 2**).

| Non-modifiable causes | Modifiable causes | | |
|---|---|---|---|
| | Hypothetical causes (no evidence) | Quantified evidenced causes | Evidenced causes not quantified |
| 2.1- Patient does not keep appointment | 1.1- Lack of staff time | 1.2- Lack of personnel | 4.1- Outdated Quality Manual |
| 2.2- Patient is not well prepared | 1.3- Lack of staff motivation | 3.1- Errors in the supply of radiopharmaceuticals | 4.2- Old equipment |
| 2.3- Patient does not remember test preparation | 1.5- Lack of training | 3.1.1- Production failures | 4.3- Variable care planning |
| 1.6- Citation errors | 1.4- Tasks shared by several physicians | 3.1.2- Order reading failures | 4.4- No quality manager |
| | | | 4.5- Lack of accessibility to the protocol |

**Table 2**. Classification of identified causes influencing the identified quality problem.

### 3.2.3.1 Criteria for assessing quality

Reliable and valid criteria were constructed to measure the initial quality and seek to document the improvement achieved. The following is a description of the selected criteria, defined criteria with their respective exceptions and clarifications based on the existing scientific evidence (**Table 3**). The following documents were used as reference:

- Law 29/2006, of July 26, 2006, on guarantees and rational use of medicines and health products. (7).

- Order SCO/2733/2007, of September 4, 2007, which approves and publishes the training program for the specialty of Radiopharmacy. (8).
- Royal Decree 673/2023, of July 18, establishing the quality and safety criteria for nuclear medicine care units. (9).
- GLUSCAN 600 MBq/mL radiopharmaceutical data sheet (3).
- SNMMI and EANM clinical practice guideline for standard and pediatric procedures for PET-CT scans with [18F]F-FDG (4).

| Criteria | Statement | Exceptions | Clarifications |
|---|---|---|---|
| **Criterion 1** | Adequacy of the posology of PET-CT scanning with [18F]F-FDG to current regulations. | Pediatric doses indicated for patients under 18 years of age. Dose for perfusion and cerebral metabolism scans. | The dosage should be according to the technical data sheet and the recommendations of the SNMMI and EANM of 3.7MBq/kg. |
| **Criterion 2** | Adequacy of the dosimetry of PET-CT scanning with [18F]F-FDG to current regulations. | Pediatric scans indicated for patients under 18 years of age. Dose for perfusion and cerebral metabolism scans. | The dosimetry of the radiopharmaceutical [18F]F-FDG should be according to the data sheet and the recommendations of the SNMMI and EANM those calculated for a dose of 3.7MBq/kg. |
| **Criterion 3** | Dispensing [18F]F-FDG must be accompanied by a physician's prescription. | No exceptions apply. | The prescription will appear on the patient's examination sheet signed by the reporting nuclear medicine physician. |
| **Criterion 4** | The supplier sends the radiopharmaceutical as requested. | No exceptions apply. | No clarifications apply. |

| **Criterion 5** | The patient comes to the appointment well prepared. | Inpatients. | Correct preparation: fasting 8h (6h for diabetics), well hydrated and blood glucose controlled. |
|---|---|---|---|

Criteria selected to evaluate the quality of the problem "Improvement and adaptation of [18F]F-FDG PET-CT scanning protocols to current legislation".

#### 3.2.3.2 Criteria validity analysis

Criterion validity is the degree to which the chosen variable correlates with an objective, reliable reference criterion that is widely accepted as a good measure of the phenomenon of interest. The analysis of the validity of the selected criteria is shown in the following table (**Table 4**).

| Attribute | Criteria | Justification |
|---|---|---|
| **Facial validity** | All | The criteria are relevant to the problem to be evaluated and are necessary to ensure compliance with current legislation and the recommendations of the most important nuclear medicine societies. |
| **Content validity** | All | The criteria measure the level of scientific-technical quality, and provide information on compliance with the level of quality of care provided. |
| **Need or expectation satisfied** | All | To provide personalized patient care, improve safety and quality for all patients undergoing PET-CT scanning with [18F]F-FDG. |
| **Validity of the criterion** | All | At present, the evidence regarding PET-CT scanning protocols with [18F]F-FDG is established by the current regulations mentioned in section 3.2.3.1. |

Analysis of the validity of the criteria selected to evaluate the quality of the problem "Improvement and adaptation of PET-CT scanning protocols with [18F]F-FDG to current legislation".

### 3.2.4 Time frame for the extraction of cases to be evaluated

The following chronology was established to evaluate the quality or compliance with the selected criteria:

- September 2022:
  - Identification of the improvement opportunity

    - From September 1 to 30, data were collected from patients attending the nuclear medicine department of the HGU Santa Lucía to undergo a PET-CT scan with [18F]F-FDG.
- October 2022:
    - Data analysis and discussion.
    - Intervention design.
        - Meetings with the nuclear medicine and hospital radiophysics services to modify the protocol.
        - Implementation of the designed intervention.
- November 2022:
    - From November 1 to 30, data were collected from patients attending the nuclear medicine department of the HGU Santa Lucía to undergo a PET-CT scan with [18F]F-FDG.
    - Reevaluation.

#### 3.2.4.1 Source of data

- For the identification of the cases or study units, the records of the PET-CT scan sheets with [18F]F-FDG of each patient were collected.
- For obtaining order records of [18F]F-FDG orders chosen in this study on compliance with criterion 4, a review of orders stored in the radiopharmacy unit database and the radiopharmacy unit's nonconformance log was performed.

#### 3.2.4.2 Identification and sampling of study units

- Sampling frame: for all criteria the sampling frame was patients who required a PET-CT scan with [18F]F-FDG.
    - The total number of patients for the first study (September 2022) was N=430.
    - The total number of patients for the re-evaluation (November 2022) was N= 432.

- Sample size: the number of cases to be evaluated for all criteria was selected to be sufficient and to guarantee the feasibility of the study. A total of 150 records were evaluated.
- Sampling method: random.
  - In the first study, 150 PET-CT scan recordings with [18F]F-FDG were randomly and randomly selected out of 430 performed in September 2022.
  - The same selection method was used in the re-evaluation, randomly and randomly. An additional 150 PET-CT scan records with [18F]F-FDG were selected from the 432 performed in November 2022.

**3.2.4.3 Type of evaluation**

- In relation to the initiative to evaluate: internal evaluation.
- In relation to the temporary action to the evaluated action: retrospective evaluation.
- In relation to the persons responsible for extracting the data: self-evaluation and cross-evaluation.

3.2.4.4 Data Collection

Data collection was performed in a spreadsheet format database (Excel) designed for this study, with different cells for the evaluation of all criteria.

**3.2.4.5** Statistical techniques

To calculate the statistical significance of the results obtained, the z value, the standard statistical value of the normal distribution, was calculated. In addition, Microsoft Excel 365 was used for graphing and data analysis.

**3.2.4 Ethical considerations**

During the conduct of this study, confidentiality regarding the identity of patients' personal data was respected in accordance with Organic Law 3/2018 of December 5, 2018, on the Protection of Personal Data and Guarantee of Digital Rights.

## 4. RESULTS

### 4.1 Analysis and presentation of the initial evaluation data

For the first evaluation, 150 [18F]F-FDG PET-CT scan sheets were randomly selected from the 430 patients performed during September 2022. For the [18F]F-FDG order records and their incidences, the data stored in the radiopharmacy unit database and the radiopharmacy unit nonconformity log of 150 [18F]F-FDG orders were randomly collected. The degree of compliance with each criterion in this first evaluation is shown in the following table (**Table 5**).

| Criteria | Absolute number of compliances | Point estimate (%) | 95%CI | %C±IC95%. |
|---|---|---|---|---|
| **1. Adequacy of the posology of PET-CT scanning with [18F]F-FDG to current regulations.** | 23 | 15,3% | 0,06 | 15,3±0,06 |
| **2. Adequacy of the dosimetry of PET-CT scanning with [18F]F-FDG to current regulations.** | 23 | 15,3% | 0,06 | 15,3±0,06 |
| **3. Dispensing of [18F]F-FDG must be accompanied by a physician's prescription.** | 0 | 0% | 0 | 0 |
| **4. The supplier ships the radiopharmaceutical as requested.** | 147 | 98,0% | 0,02 | 98±0,02 |
| **5. The patient comes to the appointment well prepared.** | 132 | 88,0% | 0,04 | 88±0,05 |

Degree of compliance with the criteria of the initial evaluation in September 2022. *Confidence interval calculation has not been adjusted for universe size since n > 10% N. Point estimate ± 1.96*standard error for 95% confidence interval.

#### 4.1.1 Quality defect analysis and intervention prioritization

The following table (**Table 6**) shows the criteria in descending order according to the number of non-compliances. The relative frequency was calculated by dividing the absolute frequency of non-compliances of the corresponding criterion by the total number of non-compliances found. The last column shows the cumulative frequency of non-compliance.

| Criteria | Absolute number of non-compliances (Absolute Frequency) | Relative Frequency (%) | Cumulative Frequency (%) |
|---|---|---|---|
| **3. Dispensing [18F]F-FDG must be accompanied by a physician's prescription.** | 150 | 35% | 35% |
| **1. Adequacy of the posology of PET-CT scanning with [18F]F-FDG to current regulations.** | 127 | 30% | 65% |
| **2. Adequacy of the dosimetry of PET-CT scanning with [18F]F-FDG to current regulations.** | 127 | 30% | 95% |
| **5. The patient comes to the appointment well prepared.** | 18 | 4% | 99% |
| **4. The supplier ships the radiopharmaceutical as requested.** | 3 | 1% | 100% |

Absolute and relative frequency of non-compliance with the criteria in the initial evaluation of September 2022.

#### 4.1.2 Initial Pareto diagram with absolute and cumulative frequency of non-compliance

For a correct estimation of the level of compliance with the criteria to be evaluated and to be able to give real knowledge of the aspects evaluated in order to highlight the criteria most in need of improvement, the graphic representation known as the Pareto diagram has been chosen. For its elaboration, the number of non-compliances has been ordered from highest to lowest and an accumulated frequency curve expressed in percentage has been completed, with the purpose of obtaining a joint visualization of the criteria that contribute a greater number of quality defects and to be able to identify in the accumulated frequency curve which ones are to intervene in their improvement.

According to the results of the first evaluation, criterion 3 was the criterion with the least compliance, followed by criteria 1 and 2 in the same proportion. On the other hand, criteria 5 and 4 have been the criteria with the highest number of non-compliances, being criterion 4 the criterion with only 3 non-compliances. The graphical representation of the Pareto diagram of this initial evaluation is shown in the following figure (**Figure 3**).

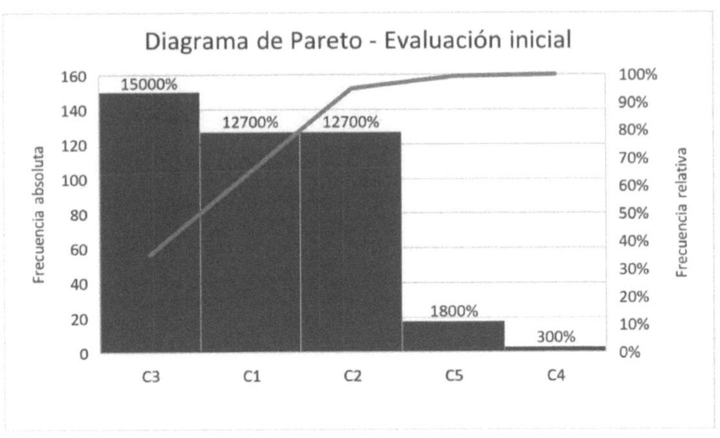


**Figure 3**. Pareto diagram: Initial evaluation September 2022.

As can be seen, criteria 3, 1 and 2 represent 95% of the quality defects or non-compliances found in this first evaluation, so the intervention will focus on them as a priority.

#### 4.1.3 Interventions to improve

For the design of the quality improvement intervention, a meeting was organized with the specialist physicians of the nuclear medicine service and the head of the radiological protection and hospital radiophysics service. At this meeting, the following strategies and activities to be carried out after the results of the first evaluation were drawn up:

- Review and documentation of current legislation.
- Brainstorming possible solutions to the quality problems encountered with the personnel involved.
- Investigation of PET-CT equipment requirements to verify the technical specifications of the equipment.
- Monitoring of equipment quality in the event of a change in dosage.
- Quality follow-up of PET-CT scans with [18F]F-FDG during the first week of the change in dosage.

### 4.1.4 Monitoring of the implementation of interventions to improve

A Gantt chart was used to monitor the interventions. This graphical representation is shown in the figure (**Figure 4**) and allows the execution of the improvement plan to be viewed and monitored chronologically. It also includes a list of interventions and those responsible for each task.

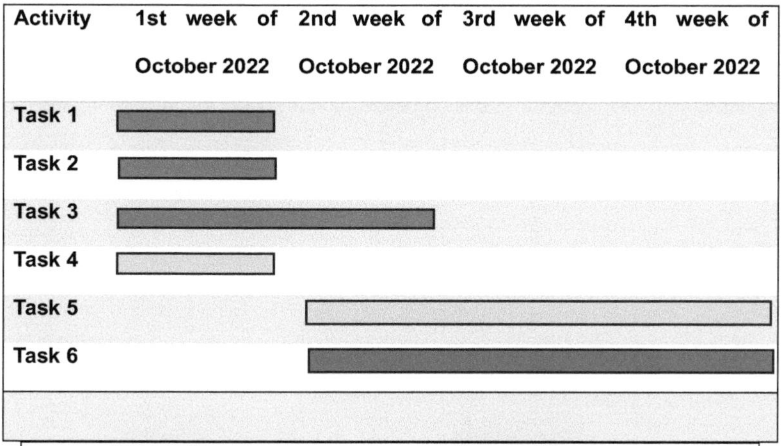

| **Activity** | **1st week of October 2022** | **2nd week of October 2022** | **3rd week of October 2022** | **4th week of October 2022** |
|---|---|---|---|---|
| **Task 1** | | | | |
| **Task 2** | | | | |
| **Task 3** | | | | |
| **Task 4** | | | | |
| **Task 5** | | | | |
| **Task 6** | | | | |

| | **Intervention** | **Responsible** |
|---|---|---|
| **Task 1.** | Review and documentation of current legislation | Ángeles García Aliaga (Radiopharmacy) |
| **Task 2.** | Brainstorming | Ángeles García Aliaga (Radiopharmacy) |
| **Task 3.** | Prescription design | Ángeles García Aliaga (Radiopharmacy) |
| **Task 4.** | Technical investigation of equipment | Hospital radiophysics |
| **Task 5.** | Monitoring of equipment | Hospital radiophysics |
| **Task 6.** | Follow-up of examinations | Nuclear medicine |

Gantt diagram: interventions to be carried out to improve and adapt PET-CT scanning protocols with [18F]F-FDG to current legislation.

## 4.2 Presentation of the data from the second evaluation of the [18F]F-FDG PET-CT scan protocol

For the second evaluation, 150 PET-CT scan sheets with [18F]F-FDG were randomly selected from the 432 patients performed during the month of November 2022. The degree of compliance with each criterion in this first evaluation is shown in the following table (**Table 7**).

| Criteria | Absolute number of compliances | Point estimate (%) | 95%CI | %C±IC95%. |
|---|---|---|---|---|
| **1. Adequacy of the posology of PET-CT scanning with [18F]F-FDG to current regulations.** | 150 | 100% | 0 | 100% |
| **2. Adequacy of the dosimetry of PET-CT scanning with [18F]F-FDG to current regulations.** | 150 | 100% | 0 | 100% |
| **3. Dispensing [18F]F-FDG must be accompanied by a physician's prescription.** | 150 | 100% | 0 | 100% |
| **4. The supplier ships the radiopharmaceutical as requested.** | 148 | 98,7% | 0,02 | 98,7±0,02 |
| **5. The patient comes to the appointment well prepared.** | 128 | 85,3% | 0,06 | 85,3±0,06 |

Degree of compliance with the criteria of the second evaluation in November 2022. *Confidence interval calculation has not been adjusted for universe size since n > 10% N. Point estimate ± 1.96*standard error for 95% confidence interval.

The second evaluation shows that criteria 1 and 2, which had a compliance rate of 15.3%, reached 100% compliance in this evaluation. This is due to the fact that in the first evaluation only patients weighing more than 90 kg met these two criteria and in the second evaluation they were met for all patients regardless of their body weight. Criterion 3 has gone from 0% compliance to 100% thanks to the inclusion of the medical prescription on the patient's examination sheet. The medical prescription system designed in this improvement cycle has also been implemented for all PET-CT scans. Finally, in criteria 4 and 5 there were no significant changes in the percentage of compliance between the two evaluations.

### 4.2.1 Non-compliance with criteria in the second evaluation

The following table (**Table 8**) shows the criteria in descending order according to the number of non-compliances. The sample size selected for the re-evaluation remained constant at 150 records for all criteria.

| Criteria | Absolute number of non-compliances (Absolute Frequency) | Relative Frequency (%) | Cumulative Frequency (%) |
|---|---|---|---|
| **5. The patient comes to the appointment well prepared.** | 22 | 92% | 92% |
| **4. The supplier ships the radiopharmaceutical as requested.** | 2 | 8% | 100% |
| **1. Adequacy of the posology of PET-CT scanning with [18F]F-FDG to current regulations.** | 0 | 0% | 100% |
| **2. Adequacy of the dosimetry of PET-CT scanning with [18F]F-FDG to current regulations.** | 0 | 0% | 100% |
| **3. Dispensing of [18F]F-FDG must be accompanied by a physician's prescription.** | 0 | 0% | 100% |

Absolute and relative frequency of non-compliance with the criteria in the second evaluation in November 2022.

The second evaluation shows that once the [18F]F-FDG PET-CT scanning protocol has been improved to comply with current regulations, criterion 5 (the patient comes to the appointment well prepared) with 92% non-compliance, is the criterion with the greatest room for improvement and may be the next target for another evaluation.

### 4.2.2 Pareto diagram of the second evaluation with absolute and cumulative frequency of non-compliance.

According to the results of the second evaluation it can be observed that criterion 5 was the criterion with the lowest compliance followed by criterion 4. The graphical representation of the Pareto diagram of this second evaluation is shown in the following figure (**Figure 5**).

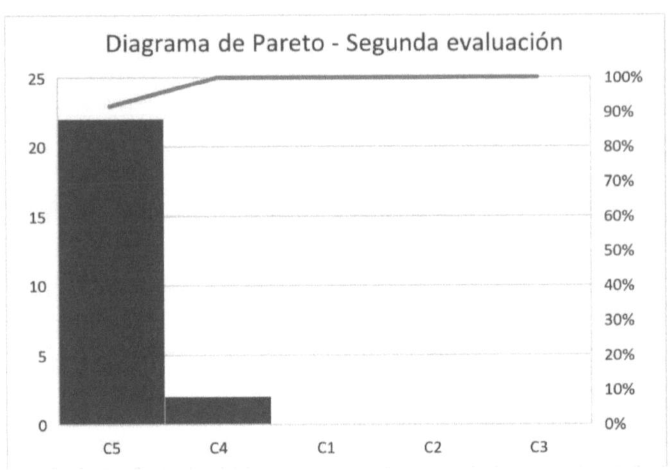


**Figure 5**. Pareto diagram: second evaluation November 2022.

### 4.3 Re-evaluation, analysis and presentation of comparative results of the two evaluations.

The presentation of the re-evaluated data is shown numerically and graphically by means of the before-after Pareto chart in order to know the level of quality achieved after the interventions implemented during the study, as well as to know if these improvements have had statistical significance compared to the result obtained in the initial evaluation that detected the quality problem to be improved.

#### 4.3.1 Joint analysis of both evaluations

The following table (**Table 9**) was prepared to determine the proportion of punctual compliance (P1 and P2) together with the 95% confidence intervals (95%CI) of the criteria evaluated and the improvement achieved (absolute and relative). These data were used to calculate the degree of statistical significance of the improvement observed according to the formula:

$$z = \frac{p_2 - p_1}{\sqrt{p(1-p)\left(\frac{1}{n_1} + \frac{1}{n_2}\right)}}$$

| Criteria | 1st Evaluation n=150 | | 2nd Evaluation n=150 | | Absolute improvement P2-P1 | Relative improvement (P2-P1) / (1-P1) | Value Z | Significance Statistics (p) |
|---|---|---|---|---|---|---|---|---|
| | P1 | ±IC95% ±IC95% ±IC95% ±IC95% ±IC95% ±IC95% ±IC95 | P2 | ±IC95% ±IC95% ±IC95% ±IC95% ±IC95% ±IC95% ±IC95 | | | | |
| **C1** | 15,3 | 0,05 | 100 | 0 | 84,7 | 100% | 15,1 | <0,001 |
| **C2** | 15,3 | 0,05 | 100 | 0 | 84,7 | 100% | 15,1 | <0,001 |
| **C3** | 0 | 0 | 100 | 0 | 100 | 100% | 17,5 | <0,001 |
| **C4** | 98,8 | 0,02 | 98,7 | 0,02 | -0,01 | -0,8% | 0,80 | 0,788 |
| **C5** | 88 | 0,04 | 85,3 | 0,06 | -2,7 | -22,5% | 0,70 | 0,758 |

**Table 9**. Joint analysis of both evaluations. P1: compliance in the first evaluation, P2: compliance in the second evaluation.

As can be seen in the table, there was an absolute improvement of 100% in criterion 3 (dispensing of [18F]F-FDG must be accompanied by a medical prescription) due to the implementation of the prescriptions on the patient's scan sheet. In addition, the modification of the posology led to an absolute improvement of 84.7% for criteria 1 (adequacy of [18F]F-FDG PET-CT scan posology to current regulations) and 2 (adequacy of [18F]F-FDG PET-CT scan dosimetry to current regulations). On the other hand, a decrease was observed in criteria 4 (the supplier sends the radiopharmaceutical as requested) and 5 (the patient comes to the appointment well prepared).

Regarding the relative improvement calculated in the table, criteria 1, 2 and 3 have improved 100% with respect to the first evaluation. Criteria 4 and 5 show a negative relative improvement of -0.8% and -22.5% respectively, which implies a worsening with respect to the first evaluation. However, it is necessary to determine whether these deteriorations are due to the intervention carried out or whether, on the contrary, they may be the result of chance. Therefore, to determine the statistical significance (p), we first obtained the Z values for each difference in proportions, using the one-tailed method because it is less restrictive.

Subsequently, from the Z values obtained by means of the formula shown above and using the values of the table of "probabilities of an extreme or tail of the normal standard curve", the results of the statistical significance column were generated on which the following analysis can be performed. Considering the risk limit is $p < 0.05$, it can be stated that the difference between the two evaluations is statistically significant for criteria 1, 2 and 3. In these cases the null hypothesis (variation due to chance) is rejected, so we can state that the improvement in these cases is real. In the case of criteria 4 and 5, by obtaining Z values of 0.80 and 0.70 with p values of 0.788 and 0.758 respectively, the null hypothesis is accepted, which would indicate that this worsening may be due to chance and not necessarily to the intervention carried out.

### 4.3.2 Before-after Pareto diagram with absolute and cumulative frequency of non-compliance

In order to graphically compare the two evaluations, a before-after Pareto diagram has been developed to highlight the improvement obtained with the intervention of this study (**Figure 6**).

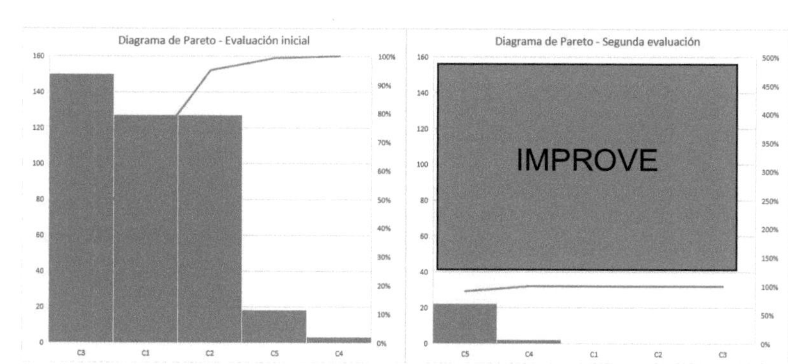


Pareto diagram before (September 2022) and after (November 2022).

This diagram shows that, on the one hand, from the 425 non-compliances identified in the first evaluation to the 24 obtained in the second evaluation, the total number of non-compliances has decreased by 401 records. This represents a very significant reduction in the number of non-compliances between the two evaluations.

On the other hand, it is evident that in the first evaluation 95% of non-compliances were represented by criteria 3, 1 and 2, while in the second evaluation, only criterion 5 accounted for 92% of the non-compliances in this evaluation. This criterion should be the object of improvement in future evaluations in order to achieve continuous quality improvement.

Finally, the height of the orange rectangle representing the improvement achieved corresponds to the difference between the total number of non-compliances between the two evaluations. In other words, 401 fewer non-compliances between the first and second evaluations, corresponding to an absolute improvement of 95%.

## 5. DISCUSSION

The intention to improve the PET-CT scanning protocol with [18F]F-FDG after reviewing the available literature and published legislation motivated the present improvement cycle, which has demonstrated that the implementation of the measures created under recommendations based on scientific evidence allows PET-CT scans with [18F]F-FDG to be performed in a safer manner for patients.

The results showed that criteria 4 (the provider sends the radiopharmaceutical as requested) and 5 (the patient comes to the appointment well prepared) did not change, since, although the relative improvement showed a worsening of these criteria in the second evaluation, this difference was not statistically significant. However, in the second evaluation, criterion 5 accounted for 95% of the noncompliances recorded and it could be interesting to carry out a cycle of improvement focused on the preparation of patients for nuclear medicine examinations. On the other hand, for criterion 4 it could be interesting to carry out annual evaluations of suppliers in order to monitor radiopharmaceutical stock-outs and gain a better understanding of the causes leading to a lack of supplies of these drugs.

With respect to criteria 1 (adequacy of the [18F]F-FDG PET-CT scan posology to current regulations) and 2 (adequacy of the [18F]F-FDG PET-CT scan dosimetry to current regulations), in the first evaluation only patients with a body mass greater than 90 kg met these criteria, this is due to the fact that the standard dosage of 370 MBq corresponds to a dose per 100 kg patient weight and according to the technical data sheet the dosage error margin may involve a margin of error of ±10%. Therefore, if we take the standard dose of 370 MBq as a reference and subtract 10%, we obtain a dose "compatible" with a body mass greater than 90 kg (333 MBq). This has meant that since the implementation in 2011 of the PET-CT protocol with [18F]F-FDG in the nuclear medicine service of the HGUSL, people with a mass of less than 90 kg have been exposed to more dosimetry in this scan than recommended by scientific societies such as the SNNMI or the EANM. The most critical case being that of patients with a body mass of 50 kg who would have received double the recommended dosimetry of [18F]F-FDG, in violation of the main principles of radiological protection.

Finally, criterion 3 (the dispensing of [18F]F-FDG must be accompanied by a medical prescription) was not met in the first evaluation and it was after the intervention (in which the prescription of PET radiopharmaceuticals was added to the PET-CT scan sheet) when this criterion began to comply with Royal Decree 673/2023, of July 18, establishing the quality and safety criteria for nuclear medicine care units. In the second evaluation, 100% compliance was obtained.

In summary, based on the identified opportunity for improvement "Improvement and adaptation of PET-CT scanning protocols with [18F]F-FDG to current legislation" the PET-CT scanning protocol with [18F]F-FDG of the nuclear medicine service of the HGUSL now complies with the provisions of: Law 29/2006, of 26 July, on guarantees and rational use of medicines and health products and Royal Decree 673/2023, of 18 July, establishing the quality and safety criteria for nuclear medicine care units. Furthermore, although this study began by focusing on the PET-CT scan protocol with [18F]F-FDG, this improvement has been extended to all PET-CT protocols performed in the nuclear medicine service of the HGUSL with different radiopharmaceuticals.

## 6. CONCLUSIONS

1. The quality problem selected after the identification of failures in the PET-CT scanning protocol with [18F]F-FDG could be subjected to an improvement cycle. This problem improved by 95% thanks to the implementation of measures and strategies carried out by the radiopharmacy unit and the nuclear medicine and hospital radiophysics services of the HGUSL.

2. The initial evaluation of compliance with the criteria established for the improvement and adaptation of [18F]F-FDG PET-CT scanning protocols to current legislation made it possible to identify the gaps in the procedure.

3. The intervention implemented as "inclusion of the prescription of PET radiopharmaceuticals in the PET-CT scan sheet" has meant the entry of the nuclear medicine service of the HGUSL into the regulatory framework of Law 29/2006, of July 26, on guarantees and rational use of drugs and health products and Royal Decree 673/2023, of July 18, establishing the quality and safety criteria for nuclear medicine care units.

4. The joint sessions carried out and the training based on scientific evidence have led to improved knowledge of PET-CT protocols, in addition to establishing a culture of quality improvement in the radiopharmacy unit and the nuclear medicine and hospital radiophysics services of the HGUSL.

5. Personalized dose adjustment according to patients' body mass is the most efficient and safe way to perform PET-CT scans with any radiopharmaceutical. In addition, customizing the dose of radiopharmaceuticals could allow the reduction and

optimization of radioactive waste generated in the radiopharmacy unit with proper planning.

## 7. BIBLIOGRAPHY .

1. Lopez-Lopez V, Robles R, Brusadin R, Lopez Conesa A, Torres J, Perez Flores D, et al. Role of 18F-FDG PET/CT vs CT-scan in patients with pulmonary metastases previously operated on for colorectal liver metastases. Br J Radiol. Jan 2018;91(1081):20170216.

Peñuelas Sánchez I. PET radiopharmaceuticals. Rev Esp Med Nucl. January 2001;20(6):477-98.

3. :: CIMA ::. GLUSCAN 600 MBQ/ML INJECTABLE SOLUTION [Internet]. [cited July 17, 2023]. Available from: https://cima.aemps.es/cima/dochtml/ft/81346/FT_81346.html

4. Vali R, Alessio A, Balza R, Borgwardt L, Bar-Sever Z, Czachowski M, et al. SNMMI Procedure Standard/EANM Practice Guideline on Pediatric 18F-FDG PET/CT for Oncology 1.0. J Nucl Med. Jan 2021;62(1):99-110.

5. Sage Journals [Internet]. [cited Aug 9, 2023]. Annals of the ICRP - Volume 38, Number 1-2, Feb 01, 2008. Available from: https://journals.sagepub.com/doi/suppl/10.1177/ANIB_38_1-2

Ministry of the Presidency. Real Decreto 1440/2010, de 5 de noviembre, por el que se aprueba el Estatuto del Consejo de Seguridad Nuclear [Internet]. Sec. 1, Royal Decree 1440/2010 Nov 22, 2010 p. 96993-7022. Available at: https://www.boe.es/eli/es/rd/2010/11/05/1440

7. BOE-A-2006-13554 Law 29/2006, of 26 July, on guarantees and rational use of medicines and health products. [Internet]. [cited August 9, 2023]. Available from: https://www.boe.es/buscar/act.php?id=BOE-A-2006-13554

Ministry of Health and Consumer Affairs. Orden SCO/2733/2007, de 4 de septiembre, por la que se aprueba y publica el programa formativo de la especialidad de Radiofarmacia

[Internet]. Sec. 3, Order SCO/2733/2007 sep 22, 2007 p. 38526-33. Available at: https://www.boe.es/eli/es/o/2007/09/04/sco2733

Ministry of Health. Real Decreto 673/2023, de 18 de julio, por el que se establecen los criterios de calidad y seguridad de las unidades asistenciales de medicina nuclear [Internet]. Sec. 1, Royal Decree 673/2023 Jul 19, 2023 p. 103988-4001. Available at: https://www.boe.es/eli/es/rd/2023/07/18/673

Printed by Books on Demand GmbH, Norderstedt / Germany